Professor John Hawk, BSc, MD, FRCP, FRACP Hon, FACD, FRSA was Emeritus professor of dermatological photobiology at the St John's Institute of Dermatology, King's College, London. He trained medically, after being very near the top in the New Zealand end-of-school national scholarship examinations, at the University of Otago, Dunedin, New Zealand and dermatologically at the St John's Institute of Dermatology, London, and Harvard Medical School, Boston, USA. He has written multiple scientific papers, chapters and books, was for years a major dermatological journal editor, has been on many journal editorial boards, lectures extensively worldwide, initiated the English care-in-the-sun campaign, leading to his invited appearance in or on many frontline newspaper articles, radio and television programmes and even magazines, helping prevent many skin cancer deaths there and worldwide thereby. He revolutionised thought and treatment concerning the 30 or so non-cancer skin diseases caused by sunlight, reclassifying them, and frequently appeared in the public media on dermatological matters. He was the British Photodermatology Group founder and president, American Photomedicine Society's co-founder, European Society for Photodermatology's co-founder, president and Honorary Life Member, member of the prestigious American Dermatological Association and European Dermatology Forum invited member, 1990 Dermatology in the year 2000 International Conference Organising Committee Chairman, 2005 EADV Congress Deputy President, 2003 – 2006 EADV Scientific Committee President, UK Advertising Standards Authority Adviser, UK Statutory Committees Adviser, including NICE on Vitamin D, President of the Edinburgh 2014 World Congress on Cancers of the Skin, International Vice-President of the Sydney 2018 World Congress on Cancers of the Skin, and was an honorary fellow of a number of international dermatological societies. He was also a consulting physician to the migration department of the New Zealand High Commission for 23 years, is a member of the invitation-only

Cook Society of academics and businessmen with Australasian and British interests and finally has been a very keen amateur cosmologist since childhood, had many sporting and cultural interests, being a life member of the Marylebone Cricket Club, member of the New Zealand Olympic Committee, former member of the Wentworth Golf Club, a keen skier and also speaking fluent French and excellent German. Professor John Hawk passed away 25th December 2022 and this book represents his final work.

John Hawk

SUNLIGHT AND US

AUSTIN MACAULEY PUBLISHERS™

LONDON · CAMBRIDGE · NEW YORK · SHARJAH

A CIP catalogue record for this title is available from the British Library.

ISBN 9781035810048 (Paperback)
ISBN 9781035810055 (Hardback)
ISBN 9781035810079 (ePub e-book)
ISBN 9781035810062 (Audiobook)

www.austinmacauley.com

First Published 2024
Austin Macauley Publishers Ltd®
1 Canada Square
Canary Wharf
London
E14 5AA

Many thanks to my dear wife, Lorna, who has encouraged this production, while also tolerating the disarray caused by this lengthy work, hopefully soon over.

Particular thanks indeed too to my extremely enterprising older son, Simon, who has moved from being a professional cricketer to a sponsorship and promotional manager for major companies to an author of his own book to an amazing cartoonist to the creator of the majority of the figures in this book, much of it contemporaneously. Well done and much thanks again, Simon.

Thanks, too, to my younger and also very successful son, Tim, who has quietly encouraged continuation of this production, even when my enthusiasm was lagging slightly.

Thanks finally to my friend, George Hayter, a former journalist and most interesting and interested chap, who advised me about the book's format and suitable publishers and similar. Many thanks again, George.

Table of Contents

Chapter One
The Origin of Sunlight

Before any full consideration of *Sunlight and Us*, it is logical first to consider how the universe itself is thought to have been formed, how our sun then developed from it and produced sunlight and how the Earth then developed, before a discussion of sunlight's interaction with us in this current era, the subject of this book.

13,800,000,000 (13.8 billion (or more precisely 13,799,000,000 +/- 21,000,000) years ago, according to and generally believed by cosmic physicists, the so-called Big Bang occurred at the beginning of and as the cause of both the space and also, perhaps unexpectedly, time of our universe. This was first suggested by Georges Lemaître in 1927, and supported thereafter by others, deduced from the rate of expansion of the universe, as assessed by Edwin Hubble's analysis of the Doppler red shift of receding galaxies shortly after in 1929, and the presence of cosmic microwave background (CMB) radiation, which has been noted throughout the universe. This presence is best explained as having developed at the same time as the Big Bang, since it could not otherwise permeate the whole of space, as it can only travel at the speed of light and therefore could not travel the billions of light years to the edges of space if it had not developed at approximately the same time. This was first suggested by Andrew McKellar in 1941 and essentially confirmed in 1964 by Nobel Prize winners Arno Penzias and Robert Woodrow Wilson. Finally, the predominance of non-heavy chemical elements (such as iron as compared with gold) points to these developing at the beginning of time since the heavier elements all require very massive stars to enable their creation through the fusion of these elements by their star's massive gravities. The Big Bang according to this hypothesis developed both inexplicably and also inexplicably quickly early on from an incomprehensibly small area of huge density and heat to form the current unspeakably vast maze of bodies of hugely variable sizes, all governed by the incontrovertible laws of physics with physical constants apparently applying throughout the whole universe (although all different in an infinite number of other similarly developing universes if such exist, as is theoretically possible according to many and mentioned below).

There are, of course, other theories concerning the origin of the universe, but none of these appears to be strongly evidence-based, but instead faith-based. This does not mean that any or all of them is wrong, especially as comprehension of the Big Bang and reason for it, even if apparently firmly evidence-based, is impossible for the human mind to comprehend in its entirety, therefore needing complete and detailed explanation, which science cannot yet reliably give. Therefore, God has generally been introduced to explain how the universe and all in it, including us, exist, but to introduce a non-communicative God, who only exceptionally rarely consults with humans, and only on an individual and therefore not independently verifiable basis, is to violate the principle of Occam's razor, proposed by a 13th-century, controversial theologian but also philosopher, which states that a second unverifiable fact should not be introduced to explain a first unverifiable fact. Thus, God may exist and have created and maintained the universe, but there is no scientific evidence of any sort yet known for it.

The creation of the universe, perhaps even one of an infinity of universes, all with different physical constants, as briefly alluded to above, occurred billions of years before the appearance of our sun, the Earth and us, the topic of this book. During these billions of years, after a very complex and variable period of development from the first picoseconds until about 370,000 years later and thereafter, swirling clouds of gas gradually coalesced into billions of galaxies of billions of stars frequently with surrounding clusters of circulating planets. All of these commenced and were completed under the influence of the very weak force of gravity, this force being determined in each case by the mass of the object. This period also included a time when the temperature and possible occurrence of water from 10,000,000 to 17,000,000 years after the occurrence of the Big Bang may in theory over this long period have permitted the temporary development, evolution and disappearance with the change of conditions of life. The force of gravity mentioned above based on an object's mass includes, for example, even our own personal feeble forces of gravity with our own miniscule masses but includes too the enormous forces associated with enormous masses.

These latter forces have led to larger stellar bodies compressing in on themselves and squashing tiny hydrogen atoms of one proton and one electron by the process known as nuclear fusion into the next larger element, helium. The latter instead has two protons and two electrons plus a neutron, and each fusion to form it releases enormous quantities of energy (according to the Einstein's equation, $E=mc2$, namely energy equals mass times the speed of light, 3,000,000 kilometres (186,000 miles) per second, multiplied by itself. This energy release includes the emission of visible light as we know it and see with, along with huge amounts of other similar and smoothly contiguous radiation but of enormously varying wavelengths and energies over a very wide total spectrum. Smaller bodies do not self-compress sufficiently to lead to nuclear fusion and radiation emission, not having enough mass and associated gravitational energy, but are subject instead to the gravity of the larger ones, either plunging into them or perhaps more fortuitously avoiding them by being gravitationally slung around them like a swirling slingshot. Such is our Earth, rotating endlessly around our sun until the sun eventually dies through using up all its energy in some 7.5 billion years' time. The light it emits replicates that emitted by all other stars, though varying in spectrum and intensity with mass, and our planet's rotation around it replicates the behaviour of all other planets, except that we by extraordinary chance are on a planet that is not too hot and not too cold to exist, along with having a suitable atmosphere and liquid water, in the so-called the Goldilocks Zone in relation to the sun, this by analogy with the perfect, not too hot and not too cold porridge Goldilocks eventually and happily found (until the bears discovered her) in the Three Bears nursery story. For interest, the Earth rotates on its own axis at about 1,670 kilometres/hour (1037 miles/hour) and orbits the sun at about 107,000 kilometres/hour (67,000 miles/hour), covering about 940.000,000 kilometres (584,000,000 miles) a year around the 150,000,000 kilometres (93,000,000 miles) distant sun.

The Big Bang hypothesis, although best at scientifically explaining the origin of the universe, and indirectly of our sun and our Earth, in an evidence-based if not humanly comprehensible way, inevitably leads to other as yet unanswered questions. For example, although time appears to have started for our universe with the Big Bang, what if anything occurred before it? Did a former universe collapse into a so-called singularity and burst through using the same matter and energy to lead to ours? How do we have our precise laws of physics and physical constants throughout our universe when an infinite number of other laws and constants are possible, none of those being likely to allow us to exist? So as mentioned before, are there an infinite number of other universes running in parallel with ours with an infinite number of different laws and constants, but not able to support us or anybody like us? And what eventually happens to our universe (and all the others if any) when all the stars run out of energy in many trillions of years' time and lose mass, energy and light? Will it all stop expanding and collapse in upon itself and start a new universe as ours apparently did? This is said not to be a very likely theory given the Second Law of Thermodynamics stating that systems with lost energy cannot reform themselves, but given that we do not understand all aspects of the universe or even its physics, it is certainly a remote possibility. In fact, recent Nobel Prize winner, Sir Roger Penrose, has apparently detected the presence of previously active, now effete Black Holes detectable in our universe, denying the Second Law of Thermodynamics and providing evidence that they are part of a previous dead universe existing where ours now is, strongly suggesting that our universe will indeed collapse in on itself in due course, perhaps therefore passing through the Big Bang singularity to repeat itself, and continue doing so for ever....

But the most massive question of all is why does anything exist at all rather than just nothing (although what would nothing be like?) And if the answer is because God devised it all, where did He come from, why did He make it so complex and why does He not explain it all to us?

There is no evidence-based answer to all this, a matter considered by many philosophers and scientists over millennia, so until the magic missing pieces of the jigsaw eventually appear (if they ever do), let us move on to the main subject of this book, something more but still by no means fully humanly comprehensible, *Sunlight and Us*.

Once the universe was formed, if the Big Bang theory is true, creating a likely expanding space at the same time, unless by much lesser chance it already existed as an inexplicable nothingness, if that is possible, waiting for a Big Bang to fill it up, early matter evolved, firstly unbelievably quickly in the first infinitesimal fractions of a second to form into vast numbers of subatomic particles. These then over the next seconds to days to weeks to months to years combined to form today's well-recognised matter, along with producing CMB (cosmic microwave background) as previously described. After a billion years of stabilisation, also alluded to previously, apparently in fact involving periods of darkness and transparency, along with matter and antimatter mostly cancelling each other out, with matter winning, and unconfirmed but almost certainly existing dark matter and dark energy in the still small but rapidly expanding universe, the universe became as it is today, with galaxies and stars continuing to form until this day, and spreading further and further out, arguably driven by postulated dark energy. And if there was in fact no dark matter, why would we be able to see between stars when there are so many in the sky behind each other to form a white blanket of light (my wife, Lorna's in-my-opinion valid theory!)? So, a billion years after the Big Bang, the universe became like ours today but much smaller, being steadily filled however by star after star in galaxy after galaxy, probably with the major help of the unconfirmed dark energy.

This of course does not yet mention the title of this section, The Early Days of Sunlight, since the sun had not yet formed. However, 9.2 billion years after the Big Bang, a random dust cloud, assisted mainly by the force of gravity, which appeared with the Big Bang, gradually swirled itself into a flat disc, with gravity pulling the dust into a central sun and surrounding planets, enabling this little book, *The Sun and Us*, to be written! Extremely simple life evolved within a few million years on one of these planets, Earth, located in the previously mentioned, so-called Goldilocks Zone around the sun, namely eventually not too hot and not too cold, and with a suitable atmosphere for life. The phenomenon of evolution then led to increasingly more complex life over billions of years, with very complex humans similar to us appearing about two million years ago and providing an even more definitive reason for the writing of this book.

Now after this very simplified description of the likely, early universe, let us proceed to the main part of this book!

Chapter Two
The Nature of Sunlight

The continuous wavelength-formed so-called electromagnetic spectrum of visible and other radiation energy emitted by the sun at hugely varying intensities is determined by its mass and the sun's consequent temperature, caused by its compression in upon itself. This is according to the theory of so-called black body radiation (though the sun although according to theory being a black body is clearly not black because of the radiation it emits as a consequence). All other stars emit radiation energy according to the same theory, but depending on their size and temperature, emit different intensities of the same spectrum of radiation.

Below is an indication of what a wavelength is, whether just a wavelength between the peaks or troughs as in the first diagram, such as of regularly rippling water, or rather different, an electromagnetic wavelength, as emitted by black bodies, namely stars, though not actually black as explained above. Electromagnetic energy developed immediately and incredibly in the first picosecond or so after the Big Bang, and was decreed then by the laws of physics to be always linked electric and magnetic radiation travelling in the same direction and at right angles to each other, and at the speed of light. Scottish physicist, James Clerk Maxwell elucidated this in 1865 and is considered one of the three greatest physicists of all time after Newton and Einstein, but conceivably now also Hawking. Thus, electricity flowing for example through a wire, or even as lightning, always creates a magnetic field around it, and a magnet drawing metal or a wire towards it always creates an associated electric current in the metal or wire.

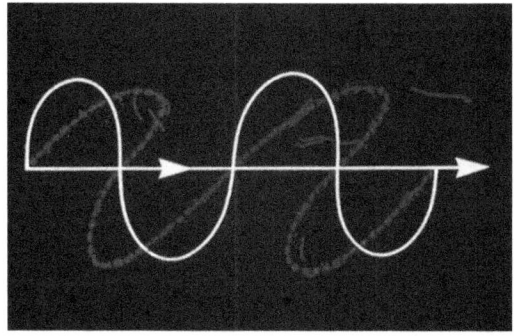

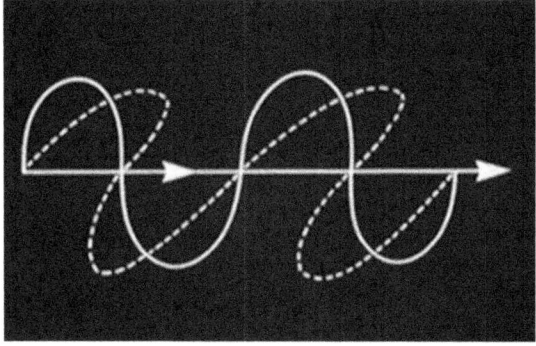

Our sun's energy of these many different wavelengths, their distribution determined by its temperature, is emitted at the enormous velocity of 3 x 108 (300,000,000) metres, (186,000 miles) per second, being difficult to measure but determined most precisely as late as 1972 using by then highly developed lasers. The frequency of the laser emission multiplied by its wavelength, now both able to be determined precisely, is indeed the speed of light, a much easier process of assessment than measuring light's speed between two points, used in initial attempted but inaccurate determinations. The exact relative intensities of these wavelengths for any body, from our bodies as an extremely small example to the sun, and to larger stars, is determined only by the temperature of the body. The emission spectrum for our sun, with surface temperature approximately 6000 degrees Kelvin, and also for bodies, usually stars, of other temperatures, is shown below, as assessed originally by 1918 Nobel Prize-winning physicist Max Planck. The visible light wavelengths are shown in their appropriate colours. The wavelengths present at the Earth's surface are more specifically marked, to indicate the ultraviolet radiation (280 – 400 nm) visible light (400 –800nm) and infrared wavelengths.

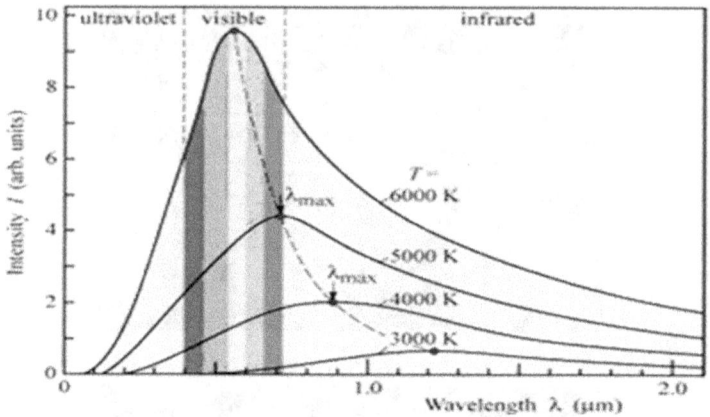

Max Plank's diagram of the electromagnetic emission spectra of so-called black bodies, in this case stars, of varying temperatures, dependent on their masses.

Such radiation provides us with the essential heat and light we need to live, as well as delivering potentially damaging, though also helpful in small amounts, ultraviolet wavelengths (UVR). Those purely UVR wavelengths reaching the Earth's surface are shown more clearly in the diagram below.

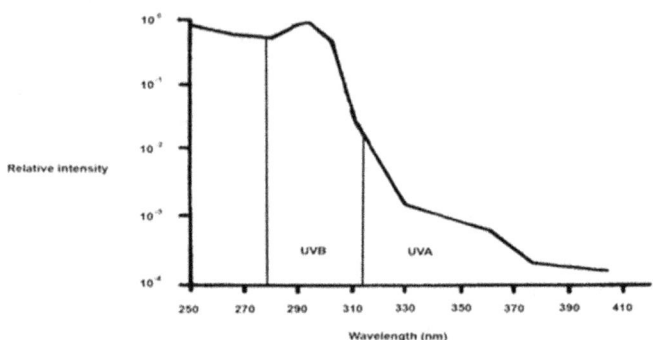

There are hugely differing effects on exposed physical matter depending on the radiation wavelength, which determines how the radiation is absorbed, if at all, by molecules in the physical matter, and in particular living tissues. Those living tissues involved are particularly the eye and the skin, both of which are susceptible to UVR injury, though both also respond usefully, the eye enabling vision and the skin absorbing warmth and producing Vitamin D (as well as other advantages, discussed later). In addition, there are multiple other solar

16

wavelengths, such as cosmic rays, gamma rays, X-rays and radio-frequency rays, but these are all too small in amount at the Earth's surface or too low in energy to affect us significantly.

When these rays enter the Earth's atmosphere, they are modified in various ways. For example, visible light is scattered by atmospheric oxygen and nitrogen molecules in such a way as to abstract the longer red rays to make the sky look blue; in addition, some of the overall radiation energy is absorbed and some reflected or scattered back into space by these molecules as well as by atmospheric water vapour, dust particles and other constituents. This leads to the penetration to ground level of only about two-thirds of the solar energy arriving at the surface of the atmosphere, at which point it consists of about five percent UVR, 40 percent visible light and 55 percent of the heat-producing infrared radiation.

Solar energy has been essential for the evolution of life on the Earth's surface, though a small proportion of it may well have developed differently deep in the heat of some areas of the oceans. This sunlight energy has provided visible light for photosynthesis, once plants had evolved, the process by which such plants grow and eventually provide food for all other creatures through the food chain (Figure 1). In addition, its infrared rays have given us the warmth we need to live, while visible light is what our eyes need to see, and what drives our life's daily rhythms. Our mood and sense of well-being are also affected by visible light; deprivation of bright light can for instance cause the winter depression known as seasonal affective disorder (SAD) (see later).

Figure 1: Visible light from the sun induces photosynthesis in plants, which enables them to grow, and provide food for herbivores (plant-eaters) such as the sheep and rabbits featured above, which then provide food for carnivores (meat-eaters) and omnivores (such as humans) eating both. These essential elements then return to the soil with the death of the animal,
and the cycle repeats.

Very small amounts of UVR, as also discussed later, much less than required to produce sunburn, to small areas of surprisingly even dark or sunscreen-protected skin for up to several days a week, also helpfully promote the skin synthesis of Vitamin D, which enables effective bone growth and function, thereby preventing rickets and osteomalacia. It also appears to assist in other important bodily functions, namely diminishing constant allergic responses to multiple potentially allergy-causing environmental substances, inducing so-called bacterial defensins to protect against skin infections with the allergic response diminished and to reduce blood pressure. Vitamin D, which also comes in our diet, for example, from fish oils, butter, egg yolks and liver can sometimes provide all we need, clearly without the added sun-induced effects mentioned above, though most of us obtain our requirements from small amounts of sun exposure. Vitamin D may also be readily and effectively obtained in safe oral supplements over the counter in pharmacy outlets if wished to ensure levels are adequate. However larger amounts of UVB are very definitely responsible for most of the harmful effects associated with sun exposure, such as skin sunburn, photo-ageing and cancers, and eye-associated snow blindness, pterygia, cataracts and perhaps macular degeneration, this latter though, if it does occur from light, from blue light exposure. Further, in spite of this, UVR is sometimes used carefully and helpfully by doctors to treat certain skin conditions if nothing else is effective, although some usually minor skin damage generally also occurs during this process, which is normally also only therapeutically effective for a few weeks to months after each usually six-week or so course for a few minutes two to three times weekly.

The UVR component of sunlight is small but biologically important, consisting of the wavelengths between 100 and 400 nanometres (nm, one thousand millionth of a metre). These are further subdivided into three categories:

- UVC: 100 – 290 nm for non-precise use, including usually also for doctors, including some dermatologists, or in strictly scientific circles 100 – 280nm.
- UVB: 290 – 320 nm, or for strict scientific use, 280 – 315nm.
- UVA: 320 – 400 nm, or for strict scientific use 315 – 400nm.

UVC is completely absorbed by ozone and oxygen in the Earth's atmosphere, so the solar UVR reaching us consists only of UVB (up to about five percent) and UVA (95 percent or more); these percentages are, however, approximate and the relative amounts vary considerably with the time of day and year, latitude and other factors. Although UVB accounts for only a small proportion of the total solar UVR, it is extremely important because its wavelengths are mainly responsible for causing sunburn, photo-ageing and skin cancer. This is because they are much more efficient than UVA at being absorbed by and damaging the genetic material of living cells, namely DNA. As a result, even though UVA forms about 95 percent of the total solar UVR around midday in summer and in the tropics, it is responsible for only about 10 to 20 percent of the harmful effects of exposure at these times, being much less absorbed by and damaging to DNA. It is now certain, however, that regularly exposing one's skin to high-dose UVA from sunbeds causes damage similar to that from sunlight, while many sunbeds have emitted a great deal of UVB as well. UVB and UVA also play an important role in the development of a whole host of unpleasant skin rashes caused by the sun.

Other Sources of UVR

By far the most important source of UVR on Earth is the sun, although the radiation is also emitted artificially and mostly, but not quite always harmlessly in very small amounts by many fluorescent and other household lamps, including in particular the new reduced energy-using compact fluorescent lamps for household lighting, replacing the traditional tungsten bulbs. These and household tungsten halogen spot lamps are both potentially dangerous if used close up regularly, as they can cause sunburn after an hour or so and probably have the potential also to cause skin photo-ageing and perhaps cancer as a result of frequent use closely over many years, though lamps with double envelopes are safer. Still, it is recommended that one should not spend long periods very close to them. UVR is also emitted by lamps used medically to treat skin disease and by arc welding equipment, and they may both be important sources of exposure for people working with them. However, the now nearly obsolete incandescent bulbs and the newish, now readily available light-emitting diode lamps (LEDs) appear harmless to the skin and eyes, and should all other lamps for household use, in spite of their greater but steadily reducing cost, offset fortunately by their much greater longevity.

How UVR Levels Vary

The factor that mainly influences the intensity of terrestrial solar UVR is the height of the sun in the sky, which depends on the time of day, season of the year and terrestrial latitude, while altitude, amount of cloud cover, type of terrain and the amount of sky visible are modifying factors of rather less importance.

Time of Day

The highest levels of UVR in the UK are received in summer within the six hours encompassing the solar zenith (when the sun is at its highest point in the sky, or 13:00 and frequently elsewhere) namely between 10:00 and 16:00. During this period, the angle of the sun relative to the Earth's surface is such that sunlight has the shortest distance to travel through the atmosphere and the least chance of being absorbed or deflected by reflection back into space or scattering in transit. As a result, about one-third of the total daily solar UVR is received between 12:00 and 14:00, and three-quarters between 10:00 and 16:00. Before 9am and after 5pm, however, the UVB and skin damage risk are minimal (even if the temperature is hot!). UVA levels though remain highish throughout the day, even early and late, so UVA effects, particularly many photodermatoses if present, remain possible all day.

The levels of UVB in particular vary significantly during the day, being much more susceptible to atmospheric effects than those of UVA and visible light. Thus, UVB intensity increases and then decreases by many times between the hours of 10:00 and 16:00 in summer, and in the tropics, where the sun is higher in the sky during the day all the year round. In practical terms, therefore, this less damaging because poorly absorbed by DNA.

An easy way to remember is that if one's shadow is shorter than one's height, one's time exposed to the sun should not be without protection. Early in the morning and later in the day, when shadows are longer, much less damage occurs to unprotected skin.

Figure 2: The higher the sun is in the sky, even if the weather is by chance cold, the less atmosphere sunlight has to traverse and the less the chance of absorption or scattering of ultraviolet radiation, particularly the more dangerous UVB, leading to much increased ultraviolet effects. At the ends of the day, the opposite applies; very little UVB getting though the greater thickness of atmosphere to be traversed, even if the weather is by chance hot.

Before 9 a.m. and after 5 p.m. in summer and the tropics, The UVB relative intensity in sunlight is very low, about one eighth of the top intensity at the solar zenith, usually 1pm. UVA, much less directly damaging to the skin, but responsible for many photodermatoses, is some five times higher throughout, until it approaches UVB relative intensity between about 11:00 and 15:00.

Season

Seasonal variations in UVR intensity, particularly UVB, are most pronounced in temperate climates, and even more so at lower latitudes.

In Northern Europe, including the UK, changes of up to 25-fold occur between winter and summer, with care needed particularly between March and October, even if the weather is cloudy, cool or even cold. UVA intensity is, however, again more constant, being less susceptible to attenuation by reflection and scattering during the passage through the atmosphere. UVR levels vary much less in the tropics, being high all year round in the daytime because the sun is always high in the middle of the day.

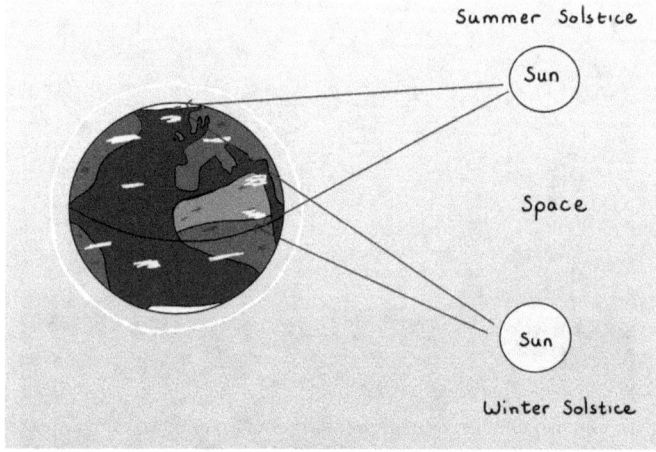

Figure 3

Geographical Latitude

The further away from the equator, the less UVR there is at any given time, the average annual exposure in Hawaii (20 degrees N) being about four times that in northern Europe (50 degrees N). This again is caused by the increased distance that UVR has to travel through the atmosphere at higher latitudes, with absorption during passage.

Altitude

For every 300 metres (or about 1,000 feet) of increase in altitude, the UVR, particularly UVB, sunburn risk increases by about four percent in the somewhat polluted northern hemisphere, though rather more rapidly in the purer southern air, again because of the shorter UVR transit distance through the atmosphere.

Figure 4: There is much less atmosphere for sunlight to traverse to reach the top of mountains, such that the UVR becomes steadily more intense as mountain altitudes increase. In addition, snow reflects UVB very easily, such that sunburn and snow blindness are easily obtained, even in mid-winter.

Cloud Cover

Clouds unless heavy may only mildly to moderately reduce the UVR intensity at ground level, having a smaller effect than on temperature, so burning is still possible on a cloudy summer's day, even if cool. This is because the water in clouds absorbs heat much better than UVR. Thus, scattered clouds in a blue sky make only a small difference to the levels of UVB, although complete light cloud cover may reduce the sunburn risk by up to 50 percent, and very heavy cloud by up to 90 percent. In other words, it is possible to burn in summer and in the tropics in cloudy, cool, dull weather. Pollution has a similar effect, again tending to reduce the effects of UVR a little, but not heat.

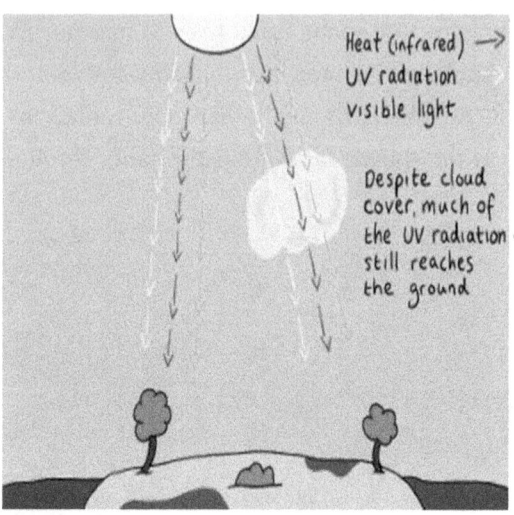

Heat (infrared) →
UV radiation
visible light

Despite cloud cover, much of the UV radiation still reaches the ground

Figure 5: UVR passing through light cloud is hardly diminished at all, even if visible light slightly diminished. Heavier cloud reduces UVR somewhat more, but heavily overcast skies are needed to ablate it nearly completely.

Wind

Wind, unless warm, has the falsely reassuring effect of reducing skin temperature so that one may feel cool even though UVB levels are unchanged. One can therefore be as badly or even worse sunburned in a breeze as without one, which is even more likely on a cloudy day, which may cause one to feel still cooler. Further, if alcohol is also being consumed, inhibition is reduced overall, along with sunburn awareness and any associated pain, making things even worse.

Window Glass

Most glass used in windows including in cars blocks UVB but not UVA nor, of course, visible light, though car windscreens do tend to give UVA protection nowadays. This means that, although glass markedly reduces the risk of sunburn, it generally does not protect against UVA-induced skin rashes and long-term ageing damage.

Surface Reflection

Some surfaces reflect UVR, allowing more to reach the skin and increasing the risk of sunburn. Thus, although grass reflects only about three percent of incident UVB, a dry, white, sandy beach reflects up to 25 percent. Further, although calm open water reflects no UVB when the sun and UVB intensity is high, rippling water and rough seas may reflect much more, perhaps up to 20 percent. This means that one will sunburn much more easily on a beach, even under a parasol, or sailing, than in the garden for example, a risk that may be increased still further by UVB scattering from the sky, as described previously, and as below. Fresh snow also reflects profuse UVB, up to 85 percent, which, together with the altitude and misleading cooling effects of wind and weather, accounts for the often severe sunburn experienced by unwary skiers, even in winter. Sunlight reflection from snow, if sunglasses are not worn, also leads to snow blindness.

Figure 6: Reflection from multiple sources, including direct rays from sun, much affecting unwary sunbather, even if sun umbrella used.

Temperature

The ambient air temperature (for example, 10°C versus 30°C) or the temperature of any water in which one may be swimming, unless on a dive at least several feet below the surface, has little influence on UVB intensity, though high temperatures do very slightly increase the damage effects.

Scattering from the Sky

UVR, particularly UVB, does not pass smoothly through the Earth's atmosphere but undergoes collisions with air molecules all the way through, much as when snooker balls collide. As a result, the rays reach the ground at all angles from the sky. Thus, if one can see lots of sky, particularly if the day is clear, there is a significant risk of UVB skin damage, even if one is protected from direct sunlight by clouds, trees, buildings or a parasol. Up to two-thirds of the UVB arrives in this way, with the rest, namely about the remaining third, arriving in a direct line from the sun.

Visible light and particularly heat are much less affected.

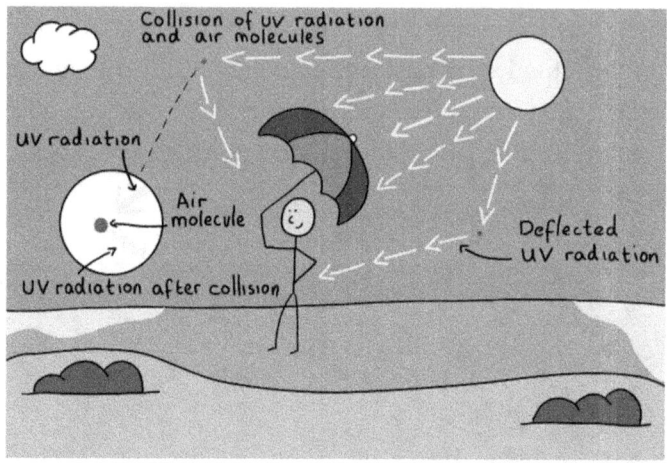

Figure 7

Ozone is a gas created from oxygen in the upper atmosphere by solar UVC radiation; this ozone then absorbs more UVC as well as some UVB, which turns the ozone back to oxygen again. In the past, there has been a delicate balance between ozone production and destruction, the absorption of all UVC and some UVB in the process preventing much of this noxious radiation from reaching the Earth. However, some decades ago, irrefutable evidence was found to indicate that this balance was gradually being upset by ozone depletion from the effects of synthetic substances such as in aerosol sprays. Fortunately, however, this occurred mostly just at times and places where the UVR intensity was weak and unimportant, such as spring in the Antarctic, though Chile and New Zealand too were a little affected with higher UVB intensities. This problem steadily worsened until very important international measures, particularly the 1997 Montreal Protocol, eventually dramatically succeeded in leading to a phasing out of the causative product use. It is now expected that atmospheric ozone levels, already much improving, will return to normal within a few further decades. If much more solar UVR had begun to reach Earth, our lives would have become seriously impaired, especially as far as food chains were concerned, with plankton in the superficial seas being severely compromised, leading to a loss of food for larger sea creatures, and then for us, and a higher risk of sun-burning, skin ageing and skin cancer for us from increased UV levels, if careful protective precautions were not taken.

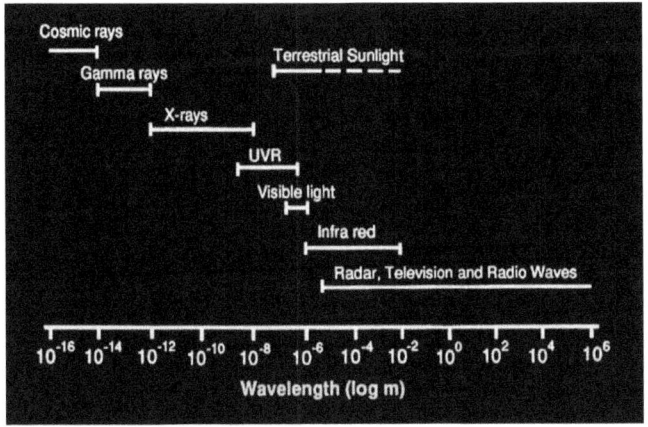

Figure 8: The whole of the solar emission spectrum, simplified, showing a huge variation in wavelength from cosmic rays (much less than a million millionth of a metre in length) to radio waves (up to a million metres in length). The shorter the wavelength, the higher the energy (Energy (E) = Planck's constant (6.63 x 10 – 34m2Kg/sec) x c (the speed of light)/the wavelength (lambda) while the radio wavelengths have very low energy. Wavelengths able to penetrate to the Earth's surface (terrestrial sunlight) labelled at the top of the diagram, are in the middle, with moderate energy, stretching from UVR to infrared (which creates heat).

It is therefore obvious that factors other than just raised UVB intensities from reduced ozone levels are leading to the rising number of human skin cancer cases during the last decades. Probably most significant is that many of us have much more time off to expose our skins to sunlight or sunbeds, the UVR from which may be similar to or occasionally even more damaging than that in sunlight. Also of relevance is that many people are living much longer, such that skin (and of course other) cancers have a longer time in which to develop. Also, a greater awareness of skin cancer may also have meant more people presenting with early lesions, some in fact able to resolve through bodily immune action if left alone, and improved diagnosis, enabling confirmation that lesions actually are cancer, are also likely to increase the apparent rate. On the other hand, the relatively recent fanaticism for sunbathing and skin tanning, very much contributing to earlier skin cancer incidence, may have significantly diminished it through much increased public health messaging, but still leading to many individuals seeing doctors about skin spots, a few cancerous, while very likely probably improving future skin cancer incidence considerably.

Chapter Three
The Sunlight and Us

a) Effects of Sunlight on Normal Skin and their Optimisation, Prevention or Treatment

Sunburn

Sunburn is caused by ultraviolet radiation in excessive amounts penetrating and being absorbed by the genetic material in skin cell nuclei, called deoxyribonucleic acid (DNA) which is the template for all cell activity but is thereby distorted, needing repair. The damage causes the inflammatory reaction known as sunburn. Because the damage is limited just to cell nuclei, where the repair process is concentrated, the inflammation has well-defined edges just to where the radiation had access, unlike other forms of inflammation, which generally have diffuse edges. The radiation causing the trouble peaks at 297 nanometres (nm) in the UVB (280 – 315nm) region, but lesser effects are shown in smoothly declining fashion on either side of this wavelength, extending even extensively but much less so into the UVA (315 – 400nm) and perhaps even minimally into the visible light region (400 – 800nm). That DNA damage does cause the reaction is essentially confirmed by the fact that the wavelengths absorbed to cause that damage are exactly the same as those shown to cause sunburn. However, many diverse chemical mediators are also recruited into the DNA-initiated repair process.

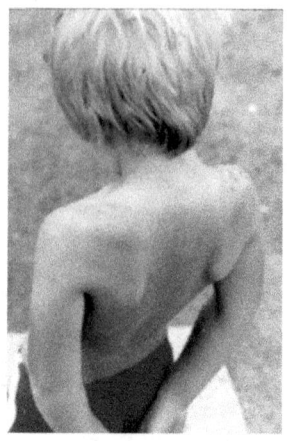

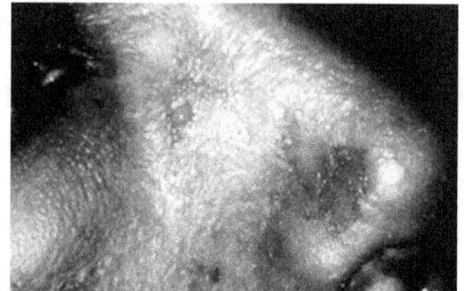

Sunburn of upper back   Sunburn of nose with blisters

Sunburn causes red, sore and occasionally very swollen and blistered skin, and if over large enough areas of the body can cause extreme generalised unwellness and even occasionally death. Sunburn is caused just by UVR as described above, and not by heat, so if cool wind as on a boat, cool water as when swimming or other cooling factors are involved but not removing the incident UVR, or particularly when there has been alcohol intake, which relieves concern about sun exposure and also the pain of sunburn, it may become very severe before noticed by the sufferer. UVR is strongest when the exposed sun is highest in a largely clear sky, such as in the middle of the day in summer or the tropics, whatever the temperature, such that the sun's rays have less absorbing atmospheric ozone and oxygen to slant through. The atmospheric molecules also bounce the particles of radiation (so-called photons) around as if they were billiard balls, so when the radiation is strong in the circumstances described above, people may be sunburned even in the shade if they can see lots of blue sky. But they will not ever be sunburned when the sun is low in the sky, even if the weather by chance is hot then. If the sky is lightly or partly cloudy, sunburn can still occur, even if more slowly. Fair skins can burn in ten to twenty minutes in strong ultraviolet conditions, darker skins more slowly.

The cellular DNA damage caused is repaired by a complex enzyme system, largely lacking in various ways in the skin disease, xeroderma pigmentosum, discussed elsewhere and often very susceptible to sunburn and always to skin cancer, but the repair system in normal skin, although excellent is not quite perfect. Therefore some minor defects persist in the long term, such that skin replacement by replication to overcome constant, normal, superficial damage, which occurs every month derived from the skin's DNA template at the bottom of its top layer, the epidermis, is gradually more impaired. This depends largely on how many sunburns occur, and even but less so for exposures causing less than sunburns. In the end, this accumulation of damage gradually leads to skin ageing along with associated deeper damage to the supporting dermis and to the increasingly impaired normal monthly skin replacement, or skin cancer if the accumulated poorly repaired cumulative damage is to important growth genes.

Pigmented skins are less susceptible to this damage, brown about five times less, black about twenty times less, as the pigment, melanin, is structured to enable absorption of much of the causative radiation before it reaches the DNA in the nuclei, before releasing it again as small amounts of non-damaging heat. Acquired tans in non-pigmented subjects also help, but by only about two times or so, and are only induced themselves as a result of the DNA damage. Very pale, non-tanning skins, normal, or particularly in totally pale albinos, are hugely at risk of sunburn, skin ageing and skin cancer if not very carefully protected from UVR exposure, as explained below.

People wanting to avoid sun-burning, skin ageing and skin cancer should stay inside except in early morning or late evening sunlight, if available, even when pleasantly warm, stay in substantial shade or cover up with protective clothing or less easily, less conveniently and more expensively, though still effectively, carefully applied high-protection sunscreens (50+) or more, replaced every two hours or so or after swimming or major perspiration. These are safe unless causing itch though harmless skin allergy; in such a case, another sunscreen with different ingredients will probably be fine.

If sunburn does inadvertently occur, ibuprofen or similar non-steroidal anti-inflammatory (NSAID) medication taken immediately can help relieve pain, and probably also speed sunburn remission, as can probably strong steroid creams, safe for a day or so, or other pain relief, or cold baths or showers until things heal over in two to three days. However, these do not decrease the DNA damage and its possible ageing and cancer outcomes.

Tanning:

Immediate Tanning

Immediate tanning tends to occur in those who already have some pigmentation, acquired or genetic, and occurs from UVR-induced immediate darkening through oxidation of the bleached pigment already present, with rapid movement of the pigment particles to cover the upper surfaces of skin nuclei. This affords some protection to nuclear DNA, most importantly of the replicating basal layer, during exposure before delayed tanning develops.

Delayed Tanning

Delayed tanning appears to be a reaction induced by the effect of excised DNA fragments during the DNA repair process. This has led to the idea of producing a skin application containing such fragments, which does seem to have been developed and shown to be effective, but not yet apparently officially marketed.

Pigment protection appears as stated above to be about five times for inherited brown skin and about twenty times for inherited black skin, meaning that skin cancer is delayed by five and twenty times respectively compared with white skin cancers, such that cancers are rare in these constitutionally dark skins. However, ageing, particularly wrinkling changes, are not similarly delayed, as long wavelength UVA radiation, not significantly cancer-producing, does penetrate very well through melanin to affect the deeper skin layers to damage the supportive dermal layer collagen, but not normally cause any form of cancer.

Skin Thickening

Damage to the DNA of the bottom epidermal layer acting as template for the monthly surface skin replacement mentioned above causes replication to cease for a day or so from the damaging influence of UVR but then induces all the epidermis but particularly the dead protective and non-damageable surface layer of the skin to thicken over several days to give further protection against continuing exposure. However, once again this only follows the initial DNA harm, which cannot be fully erased. Marked protection is possible from this added to the protection by tanning, but once again some damage will already have been done.

Both tanning and skin thickening then revert to normal over days to weeks and all protection is then lost, though the DNA damage persisting after the never perfect repair does not.

Skin Ageing

As for sunburn, UVR is the prime mediator of this process, after which multiple chemical mediators are involved in the final outcome. The main cells affected are those creating and maintaining the collagen skin-supporting fibres in the deeper dermis below the epidermis, while very probably the fibres are directly affected as well. This gradually compromises dermal structure and supporting efficiency, leading to the characteristic wrinkling of the skin of older age groups, with fair skins generally being damaged more quickly. UVA penetrates easily to this depth of the skin, whereas UVB does only slightly but may still be a major direct cause of skin ageing, although only moderately efficient at causing DNA damage to the collagen-associated cells. Because skin pigmentation is only moderately effective against UVA penetration, wrinkles develop relatively easily even in genetically dark-skinned subjects, though skin cancer caused mostly by UVB does not. Small blood vessels may be damaged by UVB too in older individuals to cause red skin blotchiness, while irregular skin pigmentation also often occurs in such people, from UVB damage to pigment cells in the epidermis. The same principles of care against UVR are essential as mentioned above, but broad spectrum sunscreen, if used, is essential to protect against both UVB and UVA. Various cosmetic procedures can reverse the appearance of wrinkling, and some creams can help a little, particularly some retinoid products, while just regular moisturising is able to plump up the skin a little to help disguise wrinkles. But it is best to avoid ageing changes by care in the sun in the first place.

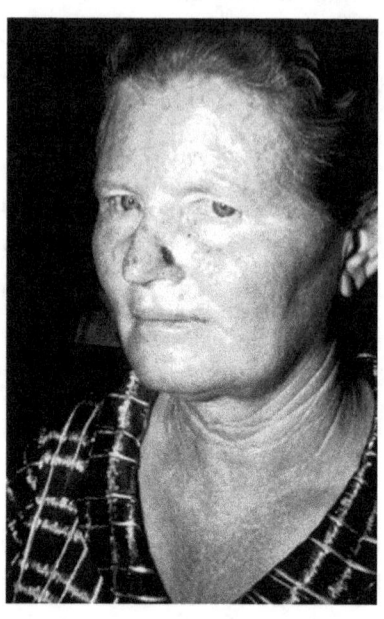

Moderate skin photo-ageing with probable basal
cell carcinoma at side of nose.

Skin Cancer

Skin cancer is caused by the outcome of accumulated UVR, mostly UVB, sun damage to the DNA areas of the bottom (basal) layer or probably also of the hair follicle cells of the skin important in its growth and replication. Damage to DNA higher up in the skin is less important as these layers are shed monthly, this replication being necessary to replace the skin damage from everyday activities, such as by constant intermittent rubbing and similar, and also more severe damage. There are various cells of different types in the basal layer, and hence various types of skin pre-cancers and cancers.

The most virulent form of skin cancer is malignant melanoma, caused from damage to the pigment-producing cells scattered evenly along the basal layer or perhaps sometimes from those in hair follicles leading to an individual's hair colour. This malignancy has almost always been rapidly fatal in the past unless excised at an early stage, usually leading to complete cure, through later spreading to many other areas of the body, thereby restricting and eventually preventing their function. Exceptionally, however, the immune system may unexpectedly recognise the malignancy and abort it, as it often does with outside invaders such as germs and viruses. Very recently though, a new class of so-called monoclonal antibody drugs can often attack and kill the melanoma cells, such as for example verumafenib and ipilimumab, with new ones constantly under development, giving a major extension of normal life, rather than rapid deterioration and death after just a few months, although the tumour does still usually (but not absolutely always) return in due course. Major sunburns in fair skin are major predisposing factor to this disease, though darker fairish skins but not usually genetically brown or black pigmented skins are still susceptible. Young adolescents and adults are definitely susceptible, unusual for most cancers, but older people too. A melanoma usually starts as a small dark, initially flat mole, less than the size of a pencil eraser, but then enlarges over just a few months to become black and also develop an irregular outline. Moles that do not change over months are not malignant, whatever their original size or shape. Some normal moles may be knocked or rubbed, often unnoticed by the patient, and become rapidly red and sore before settling back to completely normal over a couple of weeks, and these are also not malignant. Raised soft moles essentially never become malignant, though one form of melanoma, from the start, a so-called nodular melanoma, may be raised and angry looking, and steadily worsening, as distinct from soft, fleshy moles. Any doubtful changing mole not settling back to normal over a couple of weeks should be checked early by a doctor, as early treatment is always safest. Digital photos of moles on the back especially or elsewhere, checked for change every four to six months, are a good idea. A possible trap for non-experts is the very frequent appearance of totally harmless so-called seborrhoeic keratoses, usually brown, raised rough-surfaced, often very oval areas, which appear to have been stuck on to the skin like plasticine, commonest in older people and often multiple, whereas melanomas are generally single, flat and in the skin, apart from the occasional nodular form.

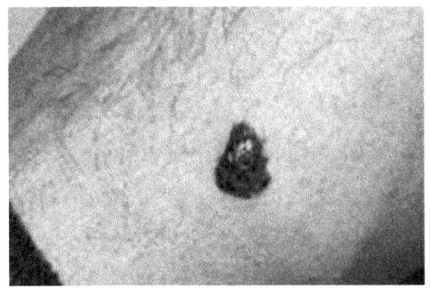

Early malignant melanoma

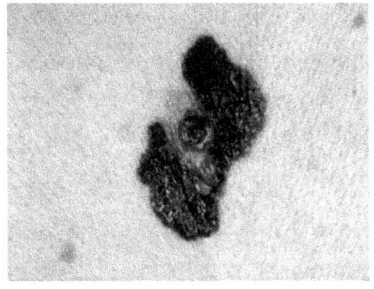

More advanced melanoma

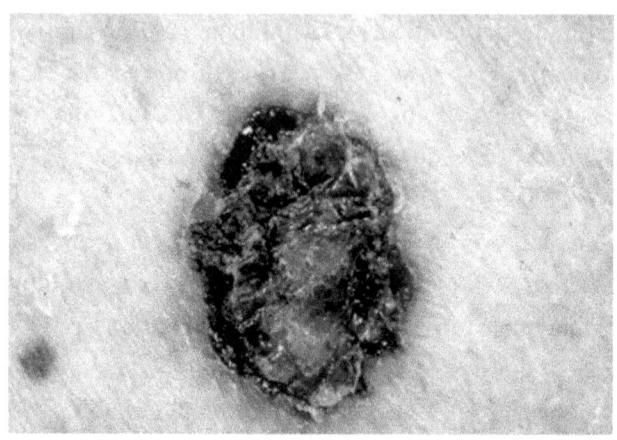

Significantly advanced melanoma

Non-melanoma skin cancers are either basal cell carcinomas (also known as rodent ulcers because they erode the skin), making up 80 percent of such cancers, and squamous cell carcinomas, with precursor cancers called actinic keratoses, harmless and easily treatable at that stage, making up the rest. There is also a very uncommon so-called Merkel cell tumour, Merkel cells being in the basal epidermal layer too and very possibly to do with the sensation of touch, as they are near the nerves which provide that sensation. Merkel cancers are all almost certainly promoted by sun exposure, and start as smooth red raised lumps, before spreading elsewhere and generally being fatal unless excised very early.

Basal cell carcinomas (rodent ulcers) most commonly affect middle-aged or older people and only grow locally (except in extraordinarily rare instances when they may sometimes spread elsewhere) but do continue to enlarge locally over months, often not bothering the patient as they are often skin-coloured and difficult to see early on. However, if they occur on areas near important sites, such as the eye or inside the ear lobe, they may erode into these structures and become very difficult to remove surgically, often leaving unpleasant scarring as a result of the removal, which they may also do on the nose. Special surgery (so-called Mohs surgery) to minimise cosmetic damage while still removing the whole tumour to avoid recurrence may be necessary. Small early lesions, particularly on the trunk, however, are easily removed by surgical excision, or if small enough and preferred, frozen with liquid nitrogen. Rodent ulcers tend to affect certain areas much more often than others, such as face below the forehead, on or in the external ears, and the trunk, though anywhere may be occasionally affected. Although caused by UVR in most instances, they do not affect all UVR-exposed skin equally, as do squamous cell carcinomas and their precursors, actinic keratoses. The reason for this is not known, and other genetic factors must be involved, as some people get them and others do not. Their totally different characteristics from squamous cell carcinomas is also not understood, but their occasional tendency to develop rudimentary hair follicle-like structures suggests they may develop from the infundibular epithelium of hair follicles, though this does not explain their different growth characteristics.

Rodent ulcers very early on may be difficult to distinguish from normal skin, but they generally grow slowly upwards and outwards, often skin-coloured, thereafter, to look somewhat abnormal. They may also often develop tiny pearly lumps within or at their edges, often with a slightly raised paler, firm-looking edge. There may often be tiny blood vessels at the edges too, sometimes with a pale, yellowish or reddish central flat area. They are often roundish in shape but may also frequently be of irregular form. Eventually through penetration to deeper, minor blood vessels, they almost always bleed slightly and become constantly covered with a blood scab. They tend not to hurt or itch, and because they grow slowly, changes may not be obvious, but they are best removed early if recognised because removal without significant cosmetic disability is then easy under local anaesthetic. If one rodent ulcer occurs, a gradual succession of new ones tends to follow over years, and the patient learns to recognise and have them treated early.

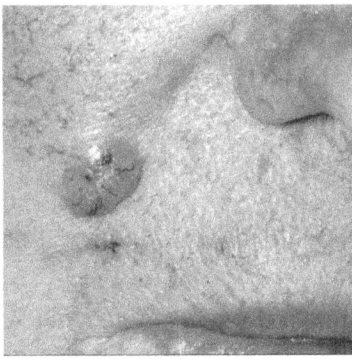

Typical fairly early (perhaps 6 – 12 months' duration) basal cell carcinoma, otherwise known as a rodent ulcer, eminently treatable by excision, or even cryotherapy on usually several occasions, without any expected recurrence, or severe outcome.

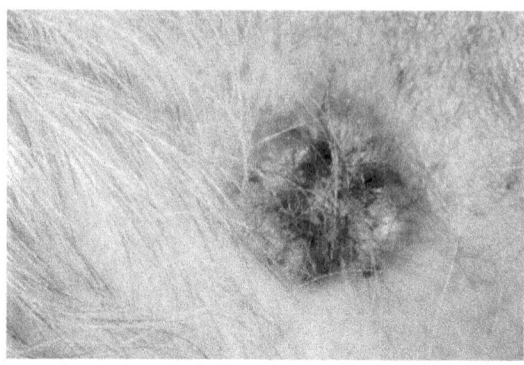

Significantly more severe basal cell carcinoma (rodent ulcer) where treatment has been delayed, and now perhaps 18 months to two years or so old, much more difficult to treat, needing surgery, best Mohs surgery to ensure margins fully excised. Complete cure without severe outcome, and usually no recurrence expected.

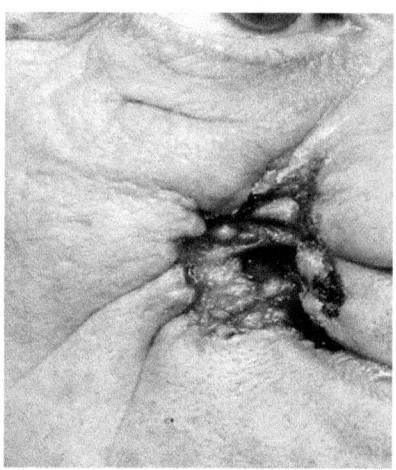

Long neglected basal cell carcinoma (rodent ulcer) maybe three – five years old, needing very difficult excision, very much preferably by Mohs surgery to ensure complete removal, but significant scarring and some facial deformity expected. However, complete cure in spite of this expected. If left further progression with erosion of deeper structures certain.

Squamous cell carcinomas are generally proceeded by rough, irregular skin-coloured, red or brownish actinic keratoses, which are commonest in pale-skinned people.

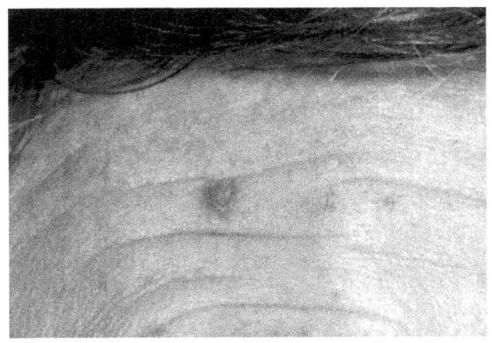 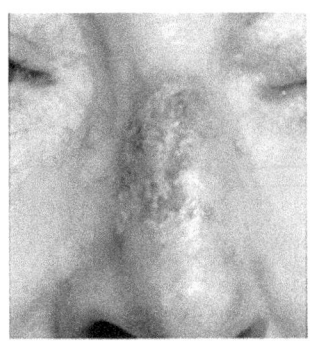

Early single actinic keratosis forehead   Much more advancedactinic keratosis nose

As a result of the warnings provided by these, care in the sun thereafter may frequently avoid the development of later squamous cell carcinomas, except much less reliably in those on immunosuppressive drugs for transplants or immunological diseases. They grow from the basal layer of the epidermis, starting as a raised red lump, which continues to enlarge, bleed and become obvious and painful. They have the potential to spread to other areas of the body but often may not until late on if left, unless the patient is on immunosuppressive drugs, when spread is more common. Surgical excision before spread is generally curative.

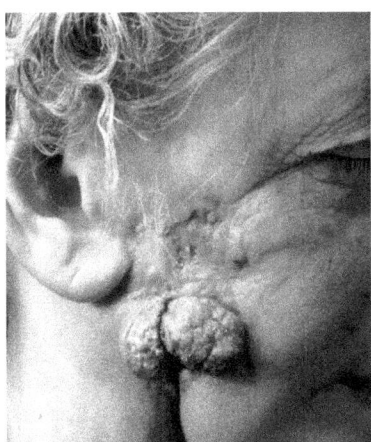

Advanced squamous cell carcinoma near right ear, long neglected, but still susceptible
to surgical removal, often curative, unless patient
also on immunosuppressive medication.

Vitamin D Formation

Vitamin D is essential to bone health and arguably to a variety of other bodily health requirements. It is available by mouth or most usually from sun exposure, though contrary to popular and media belief, as well as that of some doctors, this does not involve sun-burning levels of such exposure. Just normal life is sufficient, even surprisingly for pigmented skins, and even with careful sunscreen application to avoid sun damage, and this is confirmed by the facts that average Vitamin D levels are the same from tropical to Arctic latitudes, and that cows have normal Vitamin D levels through sun exposure through their coats (and not their bare noses) and they do not get skin cancer because of the protection from their coats. Thus, very low ultraviolet levels are all that are needed to maintain Vitamin D levels, and such low levels appear to help keep blood pressure down, to provide skin immunity to sensitisation to many environmental allergens and to lead to the production of skin defensins against external bacterial attack.

However, subjects who are bedridden or otherwise confined indoors, or always obsessively use very high protection sunscreens need oral Vitamin D supplementation, which can be taken anyway for peace of mind if preferred.

Pseudoporphyria

Pseudoporphyria is clinically exactly like porphyria cutanea tarda (PCT) as described below, but with normal porphyrin levels in blood and stool. It still seems to result from porphyrin photosensitisation to radiation which porphyrins absorb, at the junction between the UVA and visible light ranges. It is generally assisted by the ingestion of potentially photosensitising medication such as some diuretics with or without renal insufficiency with or without renal dialysis, both of which may sometimes contribute.

It also very rarely occurs with pure UVB high protection factor sunscreens along with much sun exposure. The clinical features as with actual PCT consist of fragile skin with tense blisters, particularly on the backs of the hands, which break to leave red, often infected scars which heal often to leave permanent marks, along with the presence of tiny white keratin cysts called milia, again particularly on the backs of the hands. Other features of PCT such as hair in the temporal areas and sclerodermatous changes of the chest and neck, which may on occasion occur in true PCT, are uncommon. Treatment consists of avoiding any photosensitising drugs if possible, using broad spectrum instead of just UVB sunscreens, or avoiding UVA and strong visible light exposure as far as possible. Avoiding renal dialysis, which somehow increases the likelihood of the problem is of course not possible.

Spontaneous Photo-onycholysis

Photo-onycholysis is an occasional feature of porphyria or photosensitising medication intake, but spontaneous photo-onycholysis, usually just of the finger nails, is also extremely rarely possible. This appears to be caused by excessive UVA and visible light exposure through the nails, leading to their lifting off and up at their distal, or far, ends, almost certainly promoted by normal porphyrin levels and the excessive radiation exposure. Treatment consists of avoiding such exposure, following which the nails should grow out to normal over at most the next few months.

Seasonal Affective Disorder

This condition is related to insufficient blue light reaching the retina through the eyes from sunlight, not as in the sunlight-induced conditions discussed later, related to rays from sunlight hitting the skin, not the eyes. It is well recognised and occurs to varying degrees ranging from winter blues to full blown depression, the latter much more common in those who tend to suffer depression anyway. It is commonest in northern countries with long winters with short days. It essentially never occurs in the tropics. The condition is as implied above characterised by a varying tendency to depression in winter, clearing in summer, before relapse the following winter. Occasionally, the symptoms may develop in spring instead. It appears to relate to a deficiency in production of the hormone, melatonin, secreted normally at night by the pineal gland, small and pine cone shaped, buried deep in the brain towards the rear, but linked to the eyes' perception of light by nerve connections. It is responsible for the so-called, very important, circadian rhythm of all persons, dictating bodily sleep and waking times, and other activities, such as eating and drinking times, bowel movements, urinary frequency and so on, managed by melatonin excretion by the pineal following information from the eye showing either bright or dark conditions. Probably other mood-associated hormones are involved in the process as well, such as particularly thyroid hormone and serotonin. It would seem that in the light-deprived days of winter in some individuals, too much melatonin is secreted for the person concerned by the pineal gland, and a drowsy or depressive state is induced by the melatonin until this secretion is suppressed by the arrival of brighter conditions. Daily treatment with blue light lamps during the affected months is fortunately effective, though this light should not be too bright as it may then very rarely gradually damage sight.

Other Eye Conditions Caused by Sunlight

Direct ultraviolet radiation from the sun over time may induce pterygia, an ageing condition of white thickening which gradually spreads out from the inside of one or both eyes, eventually obstructing vision as is spreads across the pupil. A pterygium may and should be removed surgically if causing reduced vision, leading to full recovery of sight. Also, cataracts are caused by UVA radiation damage occurring over many years, and macular degeneration, leading to permanent loss of direct but not peripheral vision may be related to the long-term effects of blue light.

Melasma

Melasma is generally a mildly spotty pigmented condition of the face caused by the hormone oestrogen. Sometimes severe in chloasma of pregnancy. Treatment is the VERY careful use of hydroquinone starting with one percent and going up to 10 percent if necessary.

Vitiligo

Vitiligo is a genetic depigimentation either very mild or very severe which can lead to severe burning if not protected.

b) Effects of Sunlight on Abnormal Skin and Their Treatment.

Abnormal reactions of the skin, namely skin rashes or sensations, to sunlight (the so-called photodermatoses) which only happen to some individuals in a variety of ways, are common, though most people may be unaware of that. They are distinct from the so-called normal sunlight-induced reactions, described in the last section, such as sun burning, tanning, ageing, pre-cancers and cancers, which can happen to anyone without a prior special susceptibility being necessary.

There are some 30 of these photodermatoses, or about twenty more if the originally non-sunlight-induced rashes able to be made worse by sunlight are included, which affect some 10 – 15 percent of individuals in temperate climates, and paradoxically at first sight, the reason becoming clear shortly, fewer near the sunny equator. However, most are very rare. Two of them, however, are the common ones, namely polymorphic light eruption (polymorphous in the United States), better known by sufferers in the UK as prickly heat, though heat is nothing to do with the causation, and light-exacerbated (photo-aggravated) seborrhoeic eczema (or dermatitis).

The full list of photodermatoses is listed below in Table 1.

Table 1
List of the Photodermatoses

Group A:	Group B:	Group C:	Group D:
The Immune Mediated Photodermatose[11]	*Drug and Chemical Photosensitivity.*	*The Genetic Photodermatoses*	*The Light-Exacerbated (photo-aggravated) Dermatoses (all only in some people)*
Polymorphic light eruption	**Endogenous:** **The porphyrias** a) Erythropoietic Porphyrias i) Congenital erythropoietic porphyria (Günther's disease) ii) Erythropoietic Protoporphyria b) Hepatic Porphyria i) Porphyria cutanea tarda ii) Variegate porphyria.	Xeroderma pigmentosum	Seborrhoeic eczema (dermatitis) (relatively common)

---

[1] (formerly The Idiopathic Photodermatoses, idiopathic meaning of unknown cause, until the causes have now been largely determined)

	iii) Hereditary coproporphyria		
Actinic prurigo	Exogenous a) Oral drug photosensitivity b) Topical photosensitivity	Bloom's syndrome	Atopic eczema
Solar urticaria.		Cockayne's syndrome	Lichen planus
Chronic actinic dermatitis.			Rosacea
Hydroa vacciniforme			Psoriasis (rare)
			Lupus erythematosus (common, in all forms) and many other, mostly rare, conditions.

The main two disorders mentioned above are reasonably likely to affect readers on occasion and are discussed here in detail first. The other will be considered later.

Group A – The Immune-Mediated Photodermatoses

Polymorphic Light Eruption

Polymorphic light eruption can affect anyone, but starts most commonly in the late teens or twenties, more frequently in women. It is characterised by a spotty, slightly bumpy rash of the exposed areas, but much more often just those recently exposed areas, and not the frequently exposed areas such as face and backs of hands, though these also may be affected in some people. The spots are generally red, in some patients up to two to three millimetres across and almost always itchy, in others tiny, in confluent patches and also red and itchy. The outbreak if it occurs is commoner in spring in temperate countries, and starts to appear after a few minutes to an hour or so, and fades completely in a few days or less if further sun is avoided. On sunny holidays, after often little sun exposure, it may take two to three days of such exposure to develop. It then also settles as above but may fade if exposure continues or in temperate climates as summer progresses.

It is called polymorphic though because of its variety of spots and behaviours in different people.

Diagnosis is generally obtainable from the history, and appearance of the rash, if rarely present at consultation. Lupus may perhaps need to be excluded by appropriate blood tests. Biopsy can be supportive but rarely necessary and in difficult situations, skin tests with an irradiation monochromator or solar simulator at a special hospital unit may be positive. The condition needs to be differentiated from light-exacerbated seborrhoeic eczema, as described below, by the different sites usually affected, a history of seborrhoeic eczema if present, and the appearance of the rash if present, or as described.

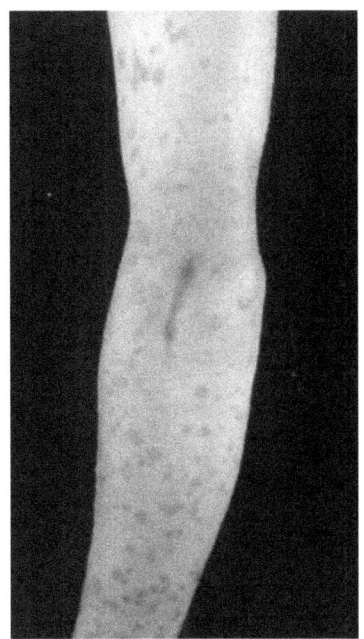

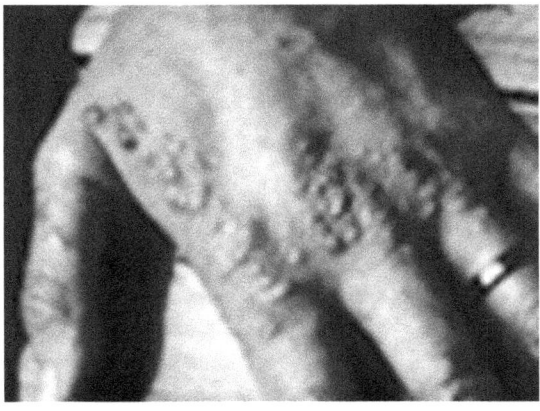

Polymorphic light eruption of arm, left, neck, middle and back of hands, right, in different patients, occurring a few hours after sun exposure, very itchy, with different appearances in different places, and lasting for up to several days until full recovery, until possible further attack.

Treatment is as a first approach preventive, consisting of avoiding sun exposure if possible and wished for, or of the careful use of highly protective, broad spectrum sunscreens against both UVB and UVA rays, on all exposed areas, preferably of SPF 50, or more if available, applied before exposure and then reapplied every couple of hours as long as exposure continues. Interestingly, this would not normally prevent Vitamin D production, even if the sunscreen is applied twice! Modern sunscreens usually work against polymorphic light eruption and, if used properly, prevent normal sunlight damage.

However, if this is not helpful or convenient, oral prednisolone steroid tablets at doses of 20 – 25 milligrams each day until the rash clears, but generally for up to two to three days to a maximum of a week from the very first sign of rash on holiday or in spring, usually turns the rash off rapidly and harmlessly, but should not be used more than twice or at most three times a year. Steroids may not be suitable for people with indigestion or worse stomach problems, those with diabetes, or those with significant psychological problems, though they often are for such short courses. A doctor needs to prescribe these.

Failing that, a course of medical low-dose ultraviolet therapy carefully administered in spring or before a holiday in graduated, low but increasing doses over four to six weeks is very frequently effective and harmless, but inconvenient because of the need to travel to the treatment centre and the length of time over which the treatment is needed. This usually works for some months. A doctor's referral for this to an appropriate hospital centre is usually needed.

In the very worst possible cases, very rare indeed, immunosuppressive therapy such as with azathioprine or cyclosporin is effective, but this approach is virtually never required.

Many other treatments have been advocated and many claimed to work without firm scientific evidence, but because of the uncertainty of occurrence of polymorphic light eruption in a wide variety of sunny conditions, the rash may not occur on any treatment, or of course on none, but any therapy that is being used at the time is then recommended. In particular, chloroquine, being frequently effective in another light-associated condition, lupus erythematosus, has frequently been used up until this day, although careful, placebo-controlled trials over 40 years ago showed chloroquine to be totally ineffective. The related hydroxychloroquine later in a similar controlled trial did surprisingly appear to have some mild effect, but the other treatments mentioned above are probably a much better approach.

The description of polymorphic light eruption and its treatment as described above does make the cause seem very difficult to analyse, but recent research suggests this is as follows. The skin's immune system turns off its immune resistance to external influences with sun exposure to reduce its likelihood of developing sun rashes (such as polymorphic light eruption (!) in which potentially allergy-causing molecules within the skin are produced, and allergic reactions to external molecules coming into contact with the skin, particularly when outside). However, in polymorphic light eruption, this immunological turning off against internal molecules damaged at any time in any way in cell nuclei, including by sunlight, is genetically reduced, in fact to no apparent advantage, such that the body continues to react allergically to these nuclear molecules, producing the rash, though continuing sun exposure does usually turn it off eventually because the genetic abnormality is not complete. That is quite a complex reason but is indeed the cause, quite commonly present in the population, for polymorphic light eruption!

Group D – The light-Exacerbated Dermatoses (included out-of-place here because so relatively common and clinically resembling polymorphic light eruption)

Light-Exacerbated Seborrhoeic Eczema

This is probably about as common as polymorphic light eruption, and is probably quite often confused with that disorder, as the time course for its appearance before and disappearance afterwards is indistinguishable. However, the rashes are completely different, but both annoyingly unsightly and itchy.

Seborrhoeic eczema, whether light-exacerbated or not, appears to be an eczematous, namely allergic reaction against penetration through the skin of molecules from an otherwise harmless, minute yeast present all over the skin of everybody, but only causing the rash in people with an inherited increased allergic tendency. The whole skin is susceptible, but actual noticeably red, itchy eczema is much more common in some areas, such as on just some or all of the following, namely the scalp (usually manifest as itching, dandruff or both), behind the ears, on the eyelids, at the sides of the nose, in the armpits and groins, in the centre of the chest and in the anal area. In some with the condition, though, sunlight exposure makes it appear from nothing or almost nothing, to become much worse, the ultraviolet radiation accentuating the allergic response, while in most patients it improves it.

The treatment for outbreaks of light-exacerbated seborrhoeic eczema is exactly the same as that for polymorphic light eruption, but to try and prevent the outbreaks, the underlying condition should also be treated routinely with a dedicated medical shampoo being used daily on the scalp to diminish the causative allergenic yeast population there, and reduce the tendency to the condition elsewhere by an internal immunological effect, along with the use of skin moisturising preparations instead of soap to wash with, thereby lessening penetration everywhere of the causative yeast molecules. If this preventive approach is insufficient to prevent sunlight-exacerbated outbreaks, the therapeutic approach mentioned earlier in the paragraph and as for polymorphic light eruption should be used.

Light-exacerbated seborrheic eczema, affecting usual sites, in normally unaffected person but flared by sunlight, although this appearance is also possible of itself in more severe instances without sunlight flaring.

These as stated are the commonest sunlight-induced conditions, and most likely to be of interest to readers, as they are mildly likely to suffer from one or other of the conditions.

The other rarer conditions in Group A are now considered too.

Actinic Prurigo

This has been known in the past as well as Hutchinson's summer prurigo, but nowadays just as actinic prurigo, actinic as previously noted meaning caused by sunlight, particularly ultraviolet radiation. Actinic prurigo appears to be a rare, prolonged, extremely itchy form of polymorphic light eruption, affecting mostly girls, and often persisting into adulthood. The reason for the prolongation of the eruption appears to be determined by a particular, rare, inherited so-called immunological haplotype, or HLA type, this being HLA-DR4, or more precisely HLA-DRB1*04, while HLA-DRB1*0407 even more emphatically predisposes to the condition. This haplotype is rare in the UK, US, Australia and New Zealand and many other countries where actinic prurigo is rare but common amongst the Native Americans of South America, where the condition is common, more so as many patients live at high altitudes near the equator, where the ultraviolet intensity is also strong. The predisposing haplotype somehow makes the condition more persistent and probably more itchy, such that the eruption is severely scratched, making it also cosmetically very unpleasant. Unlike polymorphic light eruption, it affects all exposed areas, those less frequently exposed being less affected such that there is a gradual fading up the arms and legs and sparing under any hair fringe over the forehead. It is worse during summer, and improved during winter.

The condition prevents normal behaviour during childhood because of the itch and cosmetic appearance, provoking severe social difficulties with others at school, and needs treatment as a result.

Diagnosis may be missed by physicians not experienced in the field, as the condition is very rare and may suggest severe insect bites, scabies, severe eczema or nodular prurigo, a form of non-light-induced eczema. However, particularly its characteristic, excoriated, spotty distribution over the light-exposed areas, its slow increase and then fading with the seasons in temperate areas, its usual abnormal HLA typing, and if necessary, testing in specialist centres with the irradiation monochromator or solar simulator will often if not always show abnormal results.

Treatment may be difficult, but if carefully provided by a well-informed physician is often excellent. Sunlight avoidance and the use of high protection sunscreens of SPF 50 or greater if available, may be effective in mild cases, with intermittent topical steroid ointments for short periods if needed, but resistant forms may need carefully supervised immunosuppression with cyclosporin or azathioprine, or particularly thalidomide if available. This is renowned for causing birth defects if given to pregnant mothers and extreme care should be taken to avoid this, while it can also gradually lead to an unpleasant peripheral neuropathy, particularly with tingling of the feet from nerve damage, such that regular nerve conduction studies are advised, along with stopping the drug if any peripheral neuropathy should begin to develop. Treatment with medically administered low-dose ultraviolet therapy as for polymorphic light eruption has also been reported as effective but may need to be administered in spring if no rash is then by chance present to be effective.

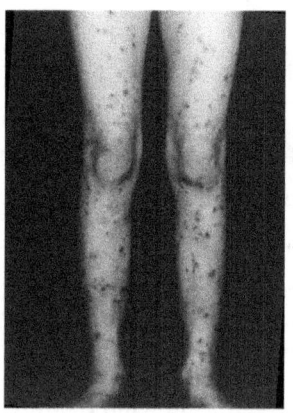

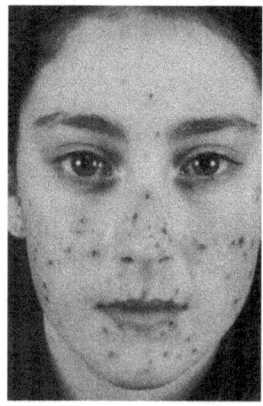 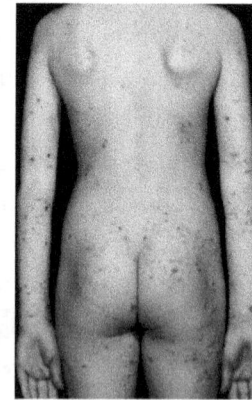

Actinic prurigo of the fronts of legs, left, face, middle and buttocks and arms, right, very itchy and cosmetically disabling, showing relative sparing of the forehead, since very often protected by a hair fringe.

Solar Urticaria

Urticaria of any type, frequently known as hives or w(h)eals in non-medical circles, is an often short-lived (over no more than a few hours, often less) itchy, red raised allergic reaction of one or multiple sites and of various sizes lasting for an hour or so but occasionally severe and widespread, or even fatal, if it worsens enough after recurring exposures to the cause. Some forms of urticaria are internal reactions against internal allergens (namely chronic urticaria) and some are reactions against external allergens or forms of energy such as heat, pressure or sunlight, the last of interest here and called solar urticaria.

Sunlight appears to induce internal allergens causing this short-lived reaction of just exposed areas, often confluent and if large areas are involved leading to fainting or even being life-threatening. Any wavelength of solar radiation from ultraviolet to visible light may be causative in different individuals, and monochromatic irradiation tests as described previously in this section reveal which.

If ultraviolet radiation alone is involved, high protection broad spectrum sunscreen use may be partly or rarely totally effective, and non-sedating antihistamines taken when exposure is expected, available over the counter in pharmacies, particularly in higher than regulation doses, may often work very well.

Failing that, difficult or very slightly risky treatments such as plasmapheresis and intravenous immunoglobulin have been used with varying but unreliable success, but nowadays with the scientific progress of medicine, six-weekly injections of the monoclonal therapy, omalizumab, also good for asthma, often work without adverse effects for all types of the condition, but a few patients remain resistant.

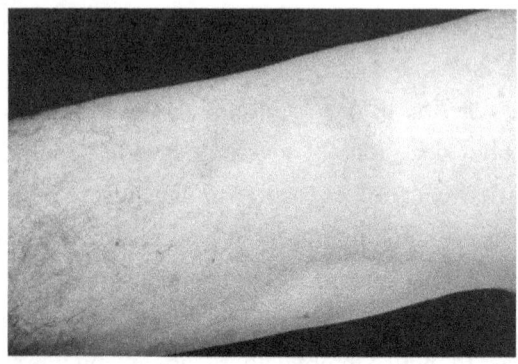

Solar urticaria of the upper arm, showing a central raised weal with surrounding red flare, induced by radiation a few minutes before, which itches severely but settles fully after an hour or so.

Chronic Actinic Dermatitis

First recognised as so-called actinic reticuloid in 1969, this condition consisted of red, extremely itchy plaques of all exposed areas, with frequent milder spread of the condition too to covered sites. Its causationy ultraviolet radiation but also often visible light wavelengths in sunlight is easily confirmed with the irradiation monochromator or solar simulator. Considered by many as a likely form of malignant lymphoma early on because of its appearance, it was later shown to be a form of eczema by the discovery that keeping patients assiduously in a dark room resulted in the resolution of the condition to an obvious eczema of the exposed areas. This was reinforced by the fact that contact eczemas with continuous exposure to the causative substance could also lead to a so-called reticuloid appearance. The condition was renamed (by me) in 1979 as chronic actinic dermatitis, a term now generally accepted for the condition. About 1:6000 subjects in temperate climates are affected by the condition, fewer in tropical areas, where the stronger sunlight suppresses the allergic tendency of all immune-mediated photodermatoses, and also other immune-mediated dermatoses, such as ordinary eczema and psoriasis.

As a form of eczema, generally known to be an allergic response to an allergen, the originating molecule for the condition was not clear, but assessment of the shape of the ultraviolet spectrum causing the disease resembled that causing sunburn, though hugely less radiation was required since the eczema occurred far more easily than sunburn. Since ultraviolet damage to cellular DNA is thought to cause sunburn, damage to DNA leading instead to allergen formation in allergic, susceptible sufferers, often already eczema patients, seems a likely cause. Treatment early on was the extremely inconvenient complete avoidance of sunlight exposure, helped a little by the poorly protective sunscreens available then. Immunosuppressive medication such as azathioprine and cyclosporin then were shown to be often very effective, along with care in the sun, though a slightly increased risk of skin cancer with sun exposure was a minor disadvantage. Very carefully administered, slowly increasing dose phototherapy to try and avoid flares was also shown to be helpful.

Nowadays, again with rapidly improving scientific advances, the injectable, monoclonal antibody medication, dipilumab, effective in atopic eczema, has also shown early promise in the treatment of chronic actinic dermatitis. It may well be a major advance in the treatment of a formerly extremely disabling condition.

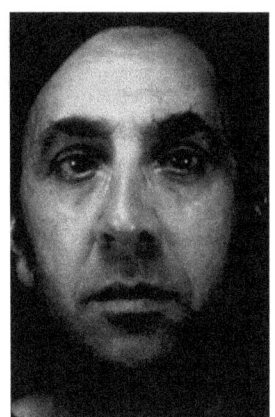 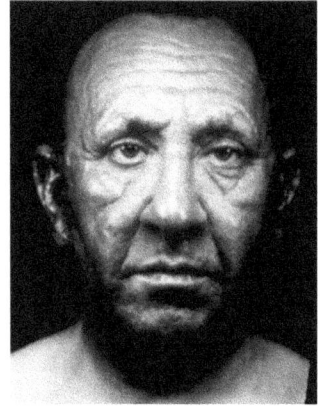

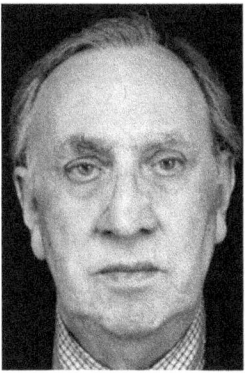

Chronic actinic dermatitis, ordinary, actinic reticuloid, same person burnt out some years later, chronic actinic dermatitis eczema all over.

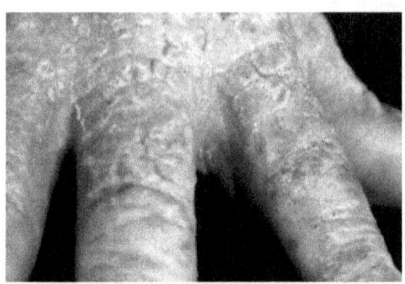

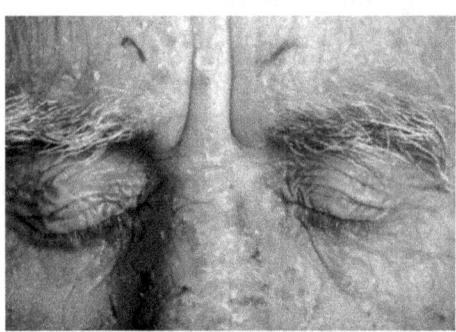
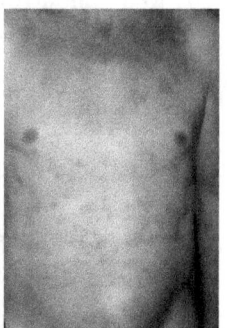

Chronic actinic dermatitis, minimally affected finger webs and eyelids because mostly light-protected.

Hydroa Vacciniforme

This very rare condition, mostly of children, is very like polymorphic light eruption in early time course, except much worse in clinical behaviour and response to treatment. It can often fade during adolescence, however.

It comes on an hour to a few hours after sun exposure, apparently UVA being most important in causing the rash, on just exposed areas but takes two to three weeks to fade, beginning with small blisters instead of small bumps as in polymorphic light eruption and then scarring badly during healing to leave persistent, unsightly pock scars. A sample of a blister with its associated skin looked at under a microscope is so characteristic as to be called pathognomonic, or diagnostic of the disorder.

The reason for the divergence of the condition from polymorphic light eruption is by no means understood, but all hydroa vacciniforme patients have the glandular fever-associated Epstein-Barr virus in the affected skin, without ever knowingly having had glandular fever, which must have an important part to play in the disease causation. Epstein-Barr virus is ubiquitous, causing no problem in most people, but does reside latently lifelong in immunological B cells and epithelium, and may promote multiple diseases and not infrequent cancers. It almost certainly has an immunological and deteriorating effect in hydroa vacciniforme, and although not reported yet in Caucasians can regularly cause hydroa vacciniforme-like T-cell lymphoma with potentially fatal consequences in children with coloured skin in China, Japan, Korea and South America, very possibly with a specific, predisposing genetic make-up, somewhat as occurs in actinic prurigo to modify it from polymorphic light eruption.

Modern high protection sunscreens can be preventive, as has been reported very carefully given phototherapy, to avoid producing the rash. The effect of immediately administered oral steroids as in polymorphic light eruption and immunosuppressive agents has not been regularly tried in hydroa vacciniforme or shown to be effective, perhaps because patients are generally young children and the disease is so rare. Overall, though, the condition is considered to be an immune-mediated condition with clinical features modified by Epstein-Barr infection.

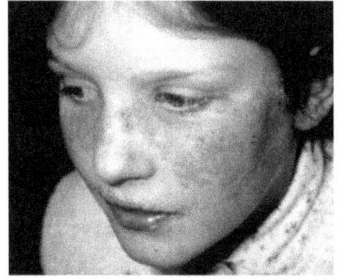

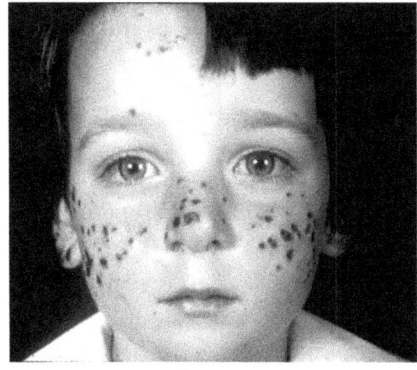 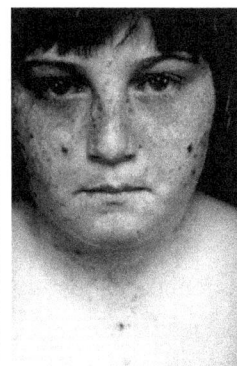

Hydroa vacciniforme, early blisters, left, then scabs, middle, then scarring face, right.

Group B: Chemical and Drug Photosensitivity – Endogenous – The Porphyrias.

The porphyrias are a complex set of largely inherited disorders of the complicated pathway leading to formation of the essential red oxygen-carrying blood pigment, haem. The commonest, a so-called cutaneous or skin porphyria, affecting very broadly 1:10,000 people, is porphyria cutanea tarda (PCT), while another rarer but still important one is erythropoietic prototoporphyria (EPP). Both are caused clinically by the fact that porphyrins, derivatives of the pathway, absorb visible light through the skin to cause damage. The others are too rare to be considered here and the two mentioned here are dealt with in some detail below, as some readers may conceivably suffer one or other.

Porphyria Cutanea Tarda

This condition is usually not directly inherited, though it may be in some 20 percent of cases, but usually sporadic, caused by a reduction to at least 20 percent of normal activity of the essential haem-pathway enzyme, uroporphyrinogen decarboxylase. The iron-dependent oxidation of uroporphyrinogen iron, being high in the liver in PCT to uroporphomethene, which then competes with uroporphyrinogen decarboxylase to diminish the concentration of itself in this condition is of prime importance in causation, though some liver-damaging personal behaviours are generally also needed to set it off. Constant, heavy alcohol intake in particular, viral liver diseases and birth control medication or hormone replacement therapy are the commonest precipitating factors in predisposed subjects. Clinical features caused by the light-absorbing porphyrins in the exposed skin consist of fragility, most notably of the backs of the hands, along with blistering, breakdown of the skin with sores, often infection and scarring, along with the characteristic formation of scattered tiny white characteristic cysts called milia, made up of compressed keratin, the substance of which the skin surface, hair and nails are made. Excess hair may also grow, particularly on the upper sides of the face. The liver is filled with porphyrin and also large amounts of iron, contributing to the occurrence of the condition, an abnormal haematochromatosis-associated iron metabolism gene often being present. The liver shows abnormal blood results and the diagnosis is made based on abnormal porphyrin findings in the urine and faeces. Treatment consists of avoiding precipitating behaviours as far as possible and either venesection every two weeks for some months until the blood iron reduces to normal, or more conveniently very low-dose amounts of the antimalarial drug, chloroquine (high doses damage the liver), which happens to attach to the abnormal porphyrins in the liver and wash them out, for some months until blood and urine tests return to normal and the clinical abnormalities do the same. All may then stay normal indefinitely, or else slow relapse often occurs, necessitating further treatment.

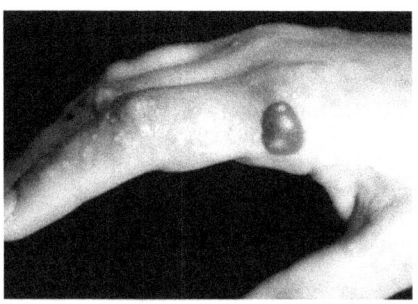

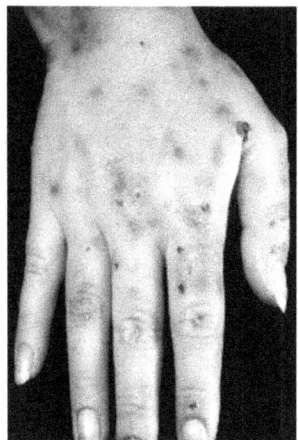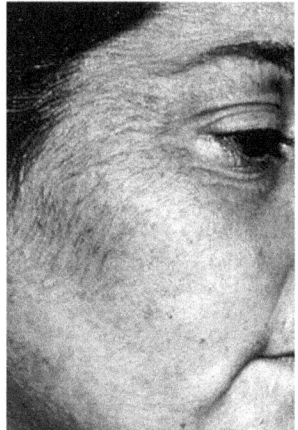

Porphyria cutanea tarda, blister (left), milia (little white spots) sores hands (middle), hair face (right)

Erythropoietic Protoporphyria

This condition is caused by a deficiency of the final enzyme in the porphyrin pathway, inserting iron to complete the creation of the oxygen-carrying red blood pigment, haem, named ferrochelatase. The causation of the deficiency is complex, requiring an autosomal dominant loss of one of the ferrochelatase genes, in combination with another gene prevalent in the general population reducing ferrochelatase activity further, and enough to cause the clinical features.

Since the enzyme deficiency is in the blood system rather than the liver, these clinical features are completely different from those of porphyria cutanea tarda. They are characterised by a painful burning or stinging sensation of the sun-exposed skin coming on in about twenty minutes to an hour or so, leading with enough exposure to redness and swelling of that skin, sometimes with skin haemorrhages and very rarely with blistering. This very painful situation settles in hours to two or two or three days or so, and leads to screaming in children, and a wish to plunge one's hands into cold water. Because the condition is rare and regularly starts in childhood, not much is known about it by non-specialists, and children may be labelled as psychologically abnormal before the true diagnosis is made. In the long term, the skin may develop a slightly shiny appearance and waxy feel, with marked radial wrinkles above the lips, along with superficial linear and tiny pock scars on the cheeks. In a very small percentage of cases, protoporphyrin overload in the liver may lead to liver failure and eventually require liver transplantation, and liver function blood tests should be undertaken annually to ensure this is not happening, although mild abnormalities are common. Any surgical operation in patients must be undertaken with great care to shield the internal organs from theatre lights, as the condition's photosensitivity may potentially lead to severe organ damage and in the worst case, patient death.

Treatment apart from sun avoidance and liberal high protection UVA sunscreen use is difficult, many potential oral therapies having been tried without any, and certainly no significant success, while medically administered ultraviolet therapy (different from the porphyria-causing light wavelengths) has also been advocated without unqualified effect. Recently, however, an analogue of melanocyte stimulating hormone, afamelanotide, which has the property of making the skin go brown, but also has immunostimulatory and anti-inflammatory effects, has been shown to increase sunlight tolerance to a reasonable degree for uncertain reasons, tanning and free radical absorption in theory seeming not to be enough. Also, for some reason, patients seem to find it better than the trial results suggest, but at least its partial efficacy does seem to have been established. The product needs to be injected subcutaneously every three months or so.

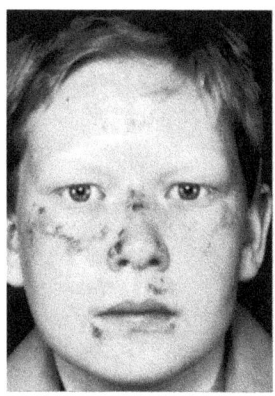

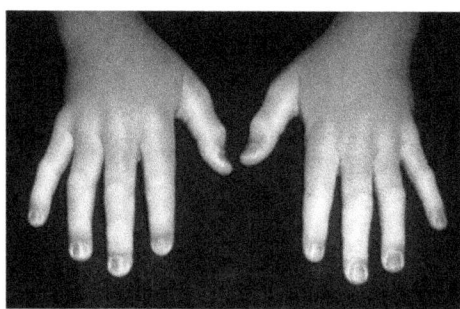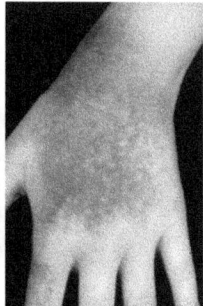

Acute, painful erythropoietic protoporphyria of face from sun exposure a few minutes earlier, left, and hands. middle, sometimes with blood spots (petechiae, right.

Typical chronic signs of erythropoietic protoporphyria, variable between patients: thickened skin over knuckles, left, radial lines above lips next, superficial waxy line scars face, third, waxy vermiculate scarring nose, last

Chemical and Drug Photosensitivity

Accidentally touching certain plants containing the photosensitiser, psoralen, in sunlight leads to bizarrely shaped redness and blistering of the affected sites following by deep tanning taking a year or so to fade fully. Responsible plants are hogweed (cow parsley) and other umbelliferae, as well as common substances such as parsley, parsnip, celery, lime and lemons, though the outdoor plants are most responsible for the skin rash, needing sunlight as well.

Multiple drugs taken by mouth nowadays can cause sunburn-like rashes in the sun, some much more often than others. Amiodarone, thiazides, tetracyclines, vemurafenib, pirfenidone and many others may be responsible but often only in some people. So many drugs may often or occasionally be responsible, normally for a sunburn-like rash, that patients should now always check for that possibility from the general practitioner, or perhaps more reliably from the dispensing pharmacist.

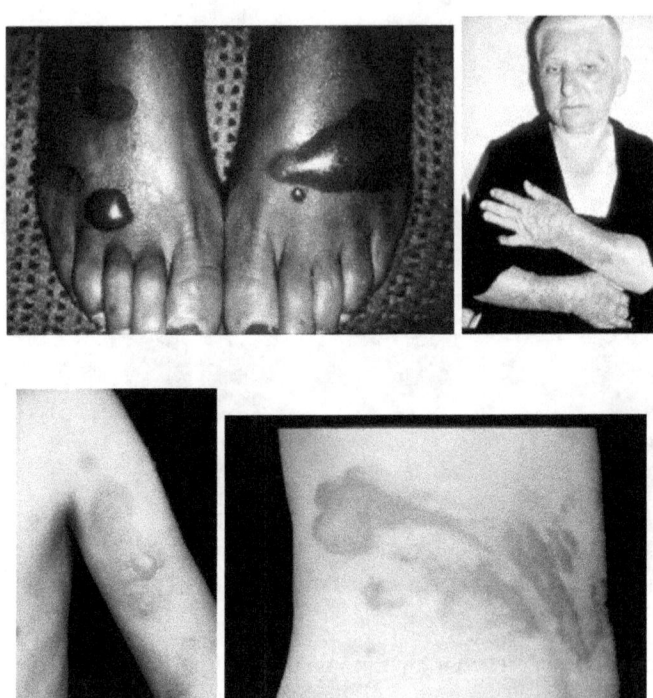

Bullous drug photo-eruption of feet, (nalidixic acid), left, generalised eruption from photosensitising drug after sun exposure next, acute eruption from recently touching psoralen-containing plant, third, months of pigmentation after acute psoralen eruption, right.

Group C
The Genetic Photodermatoses

Xeroderma pigmentosum, although rare is the commonest of these, and is caused by the inheritance of on or other of seven DNA repair enzymes or one other causing so-called xeroderma pigmentosum variant. These different enzyme deficiencies lead to the impaired repair of ultraviolet-induced sunlight damage with a variety of early symptoms and signs, mostly easy sun-burning from minimal sun exposure, along with early skin ageing and cancers, and early death therefrom, unless sun exposure is avoided. Eye problems also develop from the same cause, and a few of the same patients also develop the so-called de Sanctis-Cacchione syndrome of progressive mental deficiency. So-called T4 endonuclease V lotion delivered into the skin and to the DNA in the nucleus has been shown to enhance the reduced repair mechanisms significantly, though its use does not appear to have become widespread, with major sun-avoidance measures shared in patient groups being more common.

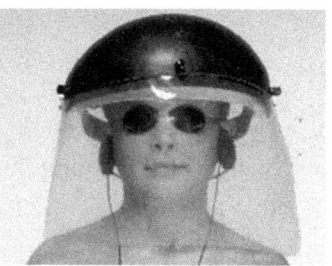

Xeroderma pigmentosum, easy sun-burning, left, careful sun protection, right.

Group D

The light-exacerbated (photo-aggravated) dermatoses.

There are a host of ordinary dermatoses that may be significantly worsened by sun exposure, generally in some but not all patients. The most serious of these is systemic lupus erythematosus, which may be made systemically worse by exposure, though light-exacerbated seborrheic eczema, mentioned in detail earlier in this book, is by far the commonest. Other disorders include lichen planus, other eczema, psoriasis rarely, and many other often rare conditions. Conventional treatment for the disorder and sun-avoidance measures to include avoidance of exposure where possible, clothing and sunscreen use are usually effective.

Chapter Four
The Future

Niels Bohr, Nobel Prize-winning Danish quantum physicist and also philosopher once said: "It is hard to make predictions, especially of the future", but in spite of that, here goes!

Three questions are of significant interest, but they may well not bother us during our lives, unless things go horribly wrong to terminate them in a moment. First how long will the Earth last and how long will we last on it? Second, how long will the sun last? Third, how long will the universe last?

It was and is difficult to be sure how the universe, sun, Earth and we started but careful scientific and archaeological studies have given us a very good idea. The sun and Earth based on scientific arguments are now about 4.6 billion years old, during most of which time some form of life, mostly very primitive, has existed, using the sun's energy to enable this. There have been several virtual extinctions of this life over this period, with further species development thereafter before the next extinction. Climate change unrelated of course to human intervention and an asteroid strike appear to have been amongst the causes responsible. Humans in their present form have only been around for some 200,000 years since our mitochondrial so-called Eve (named of course after the first biblical woman) and first genetically determined direct ancestor existed, a mere 23,000[th] of the Earth's life! In the future, whether the Earth still exists in present form, or more likely not, as discussed below, the sun will grow 10 percent hotter every billion years to become a red giant, as all stars resembling the sun do, eventually engulfing the Earth in about three billion years. The sun will then dissipate totally into mere molecules in about ten billion years, and the solar system will be no more. However, before that, in about 300 million years only (!) the Earth will no longer have the ability to sustain plant life, the basis of the food chain, because of increased temperatures and loss of carbon dioxide, their basic requirement for photosynthesis, and all life, if any still exists, will die as a result. The only place for humans to go if they have not gone already will be somewhere beyond Earth, initially perhaps to Mars as the sun expands and heats up that planet enough for life to exist, then perhaps the moons of more distant planets, then perhaps Pluto as its surface ice eventually melts with the heat from the sun's expansion. Thereafter, we must head for some more distant destination in the Milky Way, by then to include also the galaxy, Andromeda, which will have collided with the Milky Way directly in about four billion years! This should be worth watching, although the stars will probably because of distances between them mostly intermingle, though the occasional huge collision, well worth avoiding, may occasionally occur.

However, extraneous events appear likely to bring a much earlier end to our current form of life on the planet, even if not of the Earth itself, which will then carry on its merry way as described above. This early termination for us may be from another asteroid collision, a supernova explosion within a 100 light years of us, or most likely a man-made or man-associated event, such as global warming to heat us to unsustainable levels with flooding of large areas by rising seas, a nuclear explosion, a pandemic as evidenced by the Spanish influenza of 1918 – 20, the Covid-19 viral disaster of 2020 – 21, a conceivable artificial intelligence attack, or overpopulation with insufficient food to sustain us, assisted very likely by our invasion of more and more animal habitats to dispose of these habitats, useful for us, and their residents. This may all very worryingly for us happen steadily and rapidly within a mere 100 years or thereabouts, to affect most importantly our grandchildren or their children.

This last problem is very real indeed, but presumably not comprehended by those causing the problem, either because of lack of sufficient education to understand, or too great a desire for instant food, wealth or pleasure to care.

Man is however good at last-ditch self-rescues, as with the Montreal Protocol saving us from a lost ozone layer and the huge extra damaging ultraviolet influx. Can we do it again?

We do not know but we or more likely our offspring will certainly eventually find out. We have considered our past, derived by scientifically based inference, leading to our current presence here on Earth, with sunlight developing to its current status over the same period and unquestionably having been necessary for our development.

But what will happen next?

The universe itself has a number of possible outcomes but most likely is its continuing expansion, burning out of all stars, demise of black holes and utter blackness in about a trillion years to the power of twelve. However, a poorly understood force called dark energy could modify that, perhaps enabling the universe to defy the Second Law of Thermodynamics and its total lack of entropy mentioned earlier and burst out of itself (The Big Rip), (but to what end?) or collapse in on itself in The Big Crunch or follow The Big Bounce, to burst through the Big Bang singularity to start another universe all over again. Perhaps this would be along with all the multiverses mentioned earlier. Will a replica of us develop then in due course? We presumably will never know (though one, of course, can never be quite sure according to.............. not Nobel Prize winner/physicist/philosopher this time, but just me)!

It's just amazing to think that we, who think we are mostly important, are probably of no importance whatsoever compared with the vast universe which just carries on by itself regardless of us. And still, we manage to find meaning in our lives. For me, it has been the study of the sun – for what can be more fascinating than the star that is the source of life on earth, and therefore the pre-requisite for you and me.

But please always remember this source of life has great and potentially dangerous power: do make sure you protect yourself when out and about if you want to stick around to see as much of its (and our) future as you can!